A FREE GIFT F

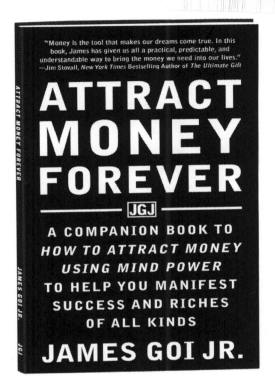

Attract Money Forever will deepen your understanding of metaphysics and mind-power principles as they relate to attracting money, manifesting abundance, and governing material reality. You'll learn how to use time-tested, time-honored, practical, and spiritual techniques to be more prosperous and improve your life in astounding and meaningful ways. Visit jamesgoijr.com/subscriber-page.html for your free download copy of this amazing book and to receive James's free monthly *Mind Power & Money Ezine*.

TEN METAPHYSICAL SECRETS OF MANIFESTING MONEY

Books by James Goi Jr.

How to Attract Money Using Mind Power

Attract Money Forever

Ten Metaphysical Secrets of Manifesting Money

Advanced Manifesting Made Easy

Aware Power Functioning

The God Function

The Supernatural Power of Thought

Ten Spiritual Secrets of Dead People

Ten Spiritual Secrets of Divine Order

Ten Spiritual Secrets of Thought Power

Self-Defense Techniques and How to Win a Street Fight

Spirituality and Metaphysics

Spiritual Power Demystified

Message from the Presence

The True Nature of Reality

Higher Consciousness

Spiritual Wisdom

The New Normal

My Song Lyrics (multiple volumes)

JGJ Thoughts, Vol. 1

Note

James continues to write new books.
To see the current list, visit his author page at Amazon.com.

TEN METAPHYSICAL SECRETS OF MANIFESTING MONEY

JGJ

SPIRITUAL INSIGHTS INTO ATTAINING PROSPERITY, RICHES, ABUNDANCE, WEALTH, AND AFFLUENCE

JAMES GOI JR.

JGJ
JAMES GOI JR.
LA MESA, CALIFORNIA

ISBN:
978-1-68347-010-6 (Trade Paperback)
978-1-68347-011-3 (Kindle)
978-1-68347-022-9 (epub)

Published by:
James Goi Jr.
P.O. Box 563
La Mesa, CA, 91944
www.jamesgoijr.com

TRADEMARK NOTICE: The Attract Money Guru™ and Books to Awaken, Uplift, and Empower™ are trademarks of James Goi Jr.

CONTENTS

Preface

PREFACE

Read this concise book over and over. Make it your daily companion. This work has a power that will become increasingly apparent over time. In writing this book, I assumed two things. One, I assumed the reader would have a great enough desire for more money to motivate him or her to read, ponder, and use this information consistently. Two, I assumed the reader would be open to metaphysical concepts. With those two foundational prerequisites satisfied, you can gain great financial benefit from what you are about to read.

I call these ten concepts *secrets* because many people simply don't know them. But throughout human history, a select few have had these insights. I present these concepts here to help you become a proficient money attractor and a more enlightened human being.

You can drastically improve your financial circumstances for the better. You can manifest the money you want. And it won't be so much a result of what you do as it will be a result of what you know and what you become.

SECRET ONE

You Already Have It

The first metaphysical secret of
manifesting money is that you
already have the money
you want to have.

Insight into Secret One

If you have conceived it, then it is already yours. It exists now, and it exists now for you. By the simple act of understanding and embracing this liberating concept, you will be much closer to realizing your financial aspirations—you will be much closer to manifesting the money you want to manifest.

Get this concept firmly within your mind: You don't have to get any money, you just need to allow to appear in tangible form the money that is already yours.

So, how is it that the money you want is already yours? Understand that your money exists before it exists. To repeat that key concept: *Your money exists before it exists.* If your money didn't exist before it existed, it would never be able to exist.

For our purposes here, there are two worlds for us to consider. First, there is the world we are most familiar with. It's the world of the seen, known material dimension. It's the world in which we live our day-to-day lives. For practical purposes, it's the only world most people are aware of or familiar with—at least while they are awake in their physical bodies.

The other world is one most people are not all that consciously aware of or familiar with. It's the unseen, unknown energy dimension.

Understand: Everything that comes into being in the world of the seen has its start in the world of the unseen. The dimension of mind, of consciousness, of energy is the breeding ground for, and the birthplace of, everything that comes into being in the material realm.

With your thoughts and feelings, you give shape in the unseen, non-material world to what eventually appears in the seen, material world. And what you create in the unseen is as real as what shows up in the seen.

The truth is that it's *more* real because what lies in the unseen is the foundation of what comes into the seen. Energy and matter are made of the same stuff. Energy and matter are interchangeable. Because of this fact, imagined circumstances are the preludes to material world circumstances.

Your concept of the money you want *is* the money you want. Your money exists before it exists. There will be a germination process, but if you stick with it, the money you have in the unseen will appear in the seen.

The first metaphysical secret of manifesting money is that you already have the money you want to have.

SECRET TWO

It Is Not Separate from You

The second metaphysical secret of manifesting money is that the money you want is not separate from you.

Insight into Secret Two

In the material world, things appear to be separate from one another. People, objects, and places appear to exist independently of other people, objects, and places. Most people believe in this illusion of separateness and never even question its validity.

In the same way, it appears that the money a person wants is separate from that person. It seems logical to conclude that if the money a person wanted were not separate from that person, he or she would already have it and thus have no reason to have to try to get it. And people are always *trying* to *get* money.

But what if you knew that the money you want is *not* separate from you? What if you knew that the money you want is, in fact, a real and actual part of you? Would that make the prospect of creating significantly greater financial resources seem more feasible?

Well, understand that your money *is* a part of you, as is everything else in creation. Your money is not at a distance from you but is instead right where you are.

Unknown by most, there is an important fact of all formed and unformed creation, and it is this: There is only One. There is only one what? There is only one everything. There is only one mind. There is only one body. There is only one place. And you *are* that mind. You *are* that body. You *are* that place.

A basic truth of the fundamental makeup of reality is that it is nonlocal. Every part of formed and unformed creation is in direct contact with every other part of formed and unformed creation at once.

When a pin drops, a vibration goes out from that pin and touches every single point of all that is. It can be a challenge to grasp this nonlocality concept. This is understandable since the concept of nonlocality seems to go against everything we think we know about reality and the structure of the material universe.

After all, some things seem to be very far apart. Look out into space, and things seem to be light years apart from each other. But it's all an illusion. The sun is no farther away from you than your hat.

And the $1,000,000, the $10,000,000, or the $100,000,000 you want is no farther away from you than the $100 in your wallet. Separateness and distance are illusions. Everything you could ever see and know is in fact—at its very core—*you*.

The second metaphysical secret of manifesting money is that the money you want is not separate from you.

SECRET THREE

It Is Not in Your Future

The third metaphysical secret of manifesting money is that the money you want is not in your future.

Insight into Secret Three

A person may accept in a metaphysical sense that they already have the money they want and that, in some abstract way, the money they want is not separate from them. But that same person might look in their wallet or at their bank statement and conclude that their money is not *actually* here *yet*.

Such a person would be under the spell of the illusion of time. And that's what time is—an *illusion*. Under the spell of this illusion, a person can assume that the things they most want are either gone now or not yet here.

This way of thinking produces a barrier between the money one wants and one actually having that money. Manifesting money is more about perception than anything else. Manifesting money is *all* about perception.

To empower yourself so you can manifest all the money you want, sharpen your perception of reality by seeing through and past the pervasive illusion of time.

Time only *seems* to exist in a material world. A material world is but a reflection of a mental construct erected in a world of energy and consciousness. And time does not exist in a world of energy and consciousness.

Some who subscribe to this outlook believe that, since time does not exist, everything has already happened and that we are coming to understand what has happened by participating in and observing a seemingly sequential stream of events. Then what about free will and our apparent ability to make choices and to make things happen? Good question. It does get convoluted. For our purposes here, let's just agree for now that we do not accurately perceive the nature of "time."

And this lack of clear, accurate perception can be counterproductive to manifesting money because while we are experiencing the illusion of time, the money we currently want never seems to quite yet be ours. But the money you want *is* yours, and it is yours *now*.

The thought that this is not true is a mental block you will do well to overcome if you want to claim your real power as the conscious creator of your financial circumstances.

So, turn away from that calendar and that clock. And take another look in your wallet and at that bank statement. Only this time, look in a new way. This time, *see* in a new way. This time, see clearly into the world of energy and consciousness so that you may know truth.

The third metaphysical secret of manifesting money is that the money you want is not in your future.

SECRET FOUR

It Is Right for You to Have It

The fourth metaphysical secret of manifesting money is that it is right for you to have the money you want.

Insight into Secret Four

Many people have misconceptions and unsettling concerns about whether it is right for people—including themselves—to have large sums of money. These people often see life as a struggle to gain a share of the limited supply of good things, including money.

The reasoning often goes that since the good things humans desire are limited, what one person has is then not available for another person to have. There's only so much to go around, it is thought, and if certain individuals have more than their "fair share" it will mean others will, as a result, not have enough.

This way of looking at things seems to make sense in a physical world, but it's just not an accurate description of how reality works, especially regarding the subject of money.

Money is just an idea. And it's a physical representation of the nonphysical concept behind it. It's a way to facilitate the exchange of goods and services. Money's potential supply is, for practical purposes, unlimited. Money, as it exists in modern society, is a renewable resource that is basically created out of thin air with a few keystrokes on a computer keyboard or by running some paper through a printing press.

Some will and do argue that our monetary system is severely flawed; regardless, it's the system we have at this time. We need to work within this present system if we are to survive and thrive financially.

This idea that it is not right for individuals to have too much money because it causes other people to have too little money is simply not valid. Actually, when a person has a lot of money, that person will tend to spend more than a person with less money and so, as a result of that spending, they will help to provide jobs and income for other people.

We do not help others to have more money by having less money ourselves. Actually, by having more money ourselves, we can help others to have more money. We do this, as was just stated above, through our spending, and we also do this through our giving and by being an example of what is possible for others.

The argument that it's wrong for you to have a lot of money because you will cause others to have less money just does not hold up under the light of logic, reason, and fact.

The fourth metaphysical secret of manifesting money is that it is right for you to have the money you want.

SECRET FIVE

You Are Worthy of Having It

The fifth metaphysical secret of manifesting money is that you are worthy of having the money you want.

Insight into Secret Five

One might not think this would be a secret and that people should know they are worthy of having the money they want. After all, why wouldn't they know this obvious fact? But experience shows that, deep down, so many people do not believe they are worthy of having the money they fantasize about having.

The origins of these feelings of unworthiness can—and often do—date back to early childhood. They often came into being during experiences we have long forgotten.

And we can also have experiences later in life that can cause or increase our feelings of unworthiness. But quite often, we do not recognize the great psychological and spiritual harm those seemingly unfortunate experiences have done to our sense of self-worth.

Aside from that, feelings of unworthiness can stem just from the fact that we are quite obviously flawed human beings who are forever making mistakes and doing things we regret and/or are embarrassed about having done. This line of thinking can naturally lead us to conclude that since we are *not* much, we don't deserve to *have* much. And that "much" can represent money or any of the other good things—both tangible and intangible—we'd choose to have and experience in our lives.

A big part of the problem is that we view ourselves as *merely* physical human beings. But physical and human are *not* all that we are. We are spiritual and divine as *well* as physical and human. In fact, our true essence *is* spiritual and divine and *not* physical and human.

Our physical expression is but fleeting and temporary, whereas our spiritual essence is lasting and permanent. We are not what we see in the mirror. We are the ones looking out from behind the eyes of our physical bodies and seeing in the mirror the present vehicles we are using to learn about ourselves.

Part of the way we are presently learning about ourselves is by expressing ourselves in modern life on planet Earth. And we are able to express ourselves so much better and more fully if we have access to all the money we care to have and can use in positive, beneficial, and healthy ways. And sure, you'll make mistakes, but always remember you are not your mistakes—you are your perfection.

The fifth metaphysical secret of manifesting money is that you are worthy of having the money you want.

SECRET SIX

A Higher Part of You Wants It

The sixth metaphysical secret of manifesting money is that a higher part of you is wanting the money you want *through* you.

Insight into Secret Six

This is true in cases where the individual is an inherently good person, meaning that he or she basically wishes the same types of things for other people as they wish for themselves.

Some people are motivated to amass fortunes so they can control and hurt other people, sometimes under the misguided impression that it will allow them to protect themselves. We are not discussing those people here.

We are referring to those who are inherently and predominantly good, caring, fair-minded human beings. There is active in good people— more so than in bad people—an unseen and more powerful part of themselves that has a hand in the affairs of their daily lives.

This other part has been called by such names as *higher self, true self, God Self, and Christ Self.* Regardless of what you call it, it is a part of you that is more godlike than human—and not human at all. It is "higher" than you in the sense of its vibration and intelligence.

Here it will be referred to as your *true self* to distinguish it from your false self, which is the physical person you currently experience yourself as being. The false physical self lives for a certain period of time. The true spiritual self is eternal. The false self is flawed. The true self is flawless. The false self often does not know what is best for itself. The true self always knows what is best for itself and for you—the false self—as well.

Human beings express desire in two basic forms: human desire and spiritual desire. Human desire is born in the mind of the false self, and the fulfillment of that desire often causes negative repercussions for the individual. Spiritual desire is born in the mind of the true self, and the fulfillment of that desire always causes positive effects for the individual.

Your true self knows that for you to become less of your false self and more of your true self, you must realize in the flesh the divine longings of the spirit.

And your true self knows what you need and what is in your highest best interests. And that is often, in part, more money than you have at a given time so that you may have the means to express yourself more fully and, therefore, more fully grow into what you were intended to be and are capable of being.

The sixth metaphysical secret of manifesting money is that a higher part of you is wanting the money you want *through* you.

SECRET SEVEN

Inspiration Beats Planning

The seventh metaphysical secret of manifesting money is that inspired action is always superior to deliberately planned action.

Insight into Secret Seven

Inspired action is divinely conceived and always appropriate and beneficial. Deliberately planned action is consciously conceived and sometimes appropriate and beneficial but sometimes inappropriate and harmful.

Inspiration is a natural transference of information, emotion, and intention from the true self of the individual and so, in effect, from the universe. The true self and the universe are, at their core, one and the same.

Planned action, which is the result of conscious-mind reasoning, springs from the inherently flawed process of human thinking.

For our purposes here, *planned action* does not strictly refer to something that is planned in advance for the purpose of being carried out at some later time. A person can have a conscious-mind thought in one moment and act on that thought in the very next moment, and such an action would fit our present definition of a planned action.

Inspired action is superior to deliberately planned action, but that's not to say that one cannot be inspired to perform certain actions in the near or distant future.

Although the resultant intention to carry out certain acts at a future time may fit the definition of a *plan*, if the idea was first divinely inspired, then it is just as good and desirable as divinely inspired acts that take place at or near the time they were inspired.

An important point here is that information regarding the true nature of circumstances and situations is not directly accessible by the human conscious mind in any way that can be deemed dependable and beyond question.

So, conscious-mind reasoning is often fundamentally flawed because the average person's conscious, intentional access to pure truth is inconsistent at best and rare at worst.

If our desire for more money is divinely motivated and sanctioned, then we will be divinely inspired to take actions that will lead to us accessing the money we are meant to have, whether or not we realize in the moment that said actions are leading to that outcome.

We can best govern our financial lives by allowing our actions to be largely motivated by our higher faculties and inclinations and by limiting the amount of misguided action we take merely as a result of our flawed conscious-mind reasoning.

The seventh metaphysical secret of manifesting money is that inspired action is always superior to deliberately planned action.

SECRET EIGHT

Be a Person Who Has It

The eighth metaphysical secret of manifesting money is that you must be a person who has the money you want to have.

Insight into Secret Eight

At any given time in your life, what you are will be the result of what you already are. You will become what you already are. You must be what you want to be before you can be what you want to be.

We're talking here about manifesting money. If you want to be, say, a person who has $100,000 in the bank, you must now be a person who has $100,000 in the bank.

The conscious, rational mind might think this instruction makes little sense. After all, one might reason, how can one be what one is not until one is that thing?

But when it comes to metaphysical and spiritual concepts, truth often runs along lines counter to conscious-mind logic and reason.

Metaphysical techniques are so underused, in part, due to people's ignorance of them and, in part, due to the fact that the logic behind them seems to the average person to be at odds with their world view and perceptions of how things work in the real world.

The fact remains that, throughout human history, many of those who have accomplished the most and acquired the most have been those who have either consciously or unconsciously acted in ways that were metaphysically and spiritually in line with, in harmony with, the way the universe works.

Physical reality springs into existence as a result of an attractive–repulsive/constructive–destructive process that is, at its core, an interplay of energy and consciousness.

Having an understanding of how this process works allows an individual to be in ways that will cause the circumstances of their lives to mold to how and what they are being.

It's an abstract concept, maybe, but a simple one to understand—though perhaps not necessarily easy to apply. Still, with some motivation and effort, it can be done.

What do you want to be? Be a person who is that. What do you want to have? Be a person who has that. Begin to think, feel, speak, and act as the person who is what you want to be and who has what you want to have.

Beyond the surface level, from deep within your core, *be* the person you want to be and who has what you want to have. Do it sincerely and with no hesitation or reservations.

The eighth metaphysical secret of manifesting money is that you must be a person who has the money you want to have.

SECRET NINE

Cooperate with the Universe

The ninth metaphysical secret of manifesting money is that you must cooperate with the universe so that it can help you get the money you want.

Insight into Secret Nine

Most people tend to work against the grain of the flow of universal energy, so most people have financial resources far below the level they would choose to have.

When you begin to consciously and methodically work *with* the grain of the flow of universal energy, you put causes into play that will allow your financial reality to come closer and closer to your financial ideal.

There's a lot of talk about the *subconscious mind* in metaphysical circles. There's also a lot of talk about the so-called *universe*. For practical purposes, there is no reason to make a distinction between the two.

Your subconscious mind has the far-reaching power it does, precisely *because* it functions in unison with the universe. By governing the functioning of your subconscious mind, you are governing the workings of the universe. The power you are capable of consciously wielding is beyond adequate description.

Since you're reading this book, you're likely no stranger to metaphysical concepts. You're likely familiar with such staple metaphysical techniques as affirmation and visualization.

You likely know of the great power of desire, belief, and expectancy. You're likely aware that manifesting is, at its core, more about your thoughts and feelings than anything else.

You likely know that a proper money mindset is an important part of being able to attract an abundant supply of money, and that developing a positive money mindset is, in part, accomplished by understanding metaphysical concepts and using metaphysical techniques.

And if you don't know these things, you can easily learn them. There is a wealth of knowledge available in books, on the Internet, and in just so many places.

Knowing the information is out there won't help you unless you find it. Finding the information won't help you unless you use it. By using even just one simple metaphysical technique diligently and purposefully—such as affirmation, for example—you could transform your financial circumstances.

Affirm now something like this: "Large sums of money continue to flow to me freely and easily." Keep affirming that, and you will, over time, see that large sums of money will continue to flow to you freely and easily.

If you are not intentionally using metaphysical techniques to attract money, you are bringing in much less money than you otherwise could be bringing in. Things work in a certain way and in no uncertain terms.

The ninth metaphysical secret of manifesting money is that you must cooperate with the universe so that it can help you get the money you want.

SECRET TEN

Spread the Good Around

The tenth metaphysical secret of
manifesting money is that you must
spread around the good that comes
to you so that you can have
the money you want.

Insight into Secret Ten

You can continue to manifest more and more money into your life. To make sure the money you manifest becomes a blessing to you, and not a curse, be willing to allow some of it to pass through you to other people.

Spending money and paying taxes is not enough. We're talking about giving. To continue to manifest money in a healthy and sustainable way, you'll need to continually allow a certain percentage of the money you receive to go freely out from you without the expectation that you will receive anything in return from the places and people you give to.

Tithing—the practice of giving ten percent of one's income to the source of one's spiritual sustenance—is a powerful and beneficial practice. But some people don't have an identifiable source of spiritual sustenance. And some just don't have enough confidence and faith to be able to freely and joyfully give away ten percent of the money they receive.

Whatever may be true for you, one thing is certain: One of the most useful practices for causing yourself to manifest more money is to consistently give away a certain percentage of all the money you receive.

You can give the money away to needy or deserving individuals or to causes you believe in and which help to make the world a better place in which to live.

If you are too afraid or selfish to give some of the money you receive back out into the universe, then you may well find that the universe does not give freely to you and that it may seem to be conspiring to take away from you even the money you already have.

Giving is not only a benevolent practice but also a financial and overall self-defense practice. The universe is always giving—and especially to the givers. But the universe can also do some taking—especially from the takers.

However much of your income you decide to give back to the source from whence it came, be sure that you do your giving with an open heart and in sincere joy for the opportunity to give and for the experience of giving. You can get, you can receive, but you must also give.

The tenth metaphysical secret of manifesting money is that you must spread around the good that comes to you so that you can have the money you want.

AFTERWORD

Thank you for reading this book. I hope you have enjoyed it. And I hope you will continue to benefit from having read it. I have benefited greatly from having read countless books over the years.

I began to find my first self-help, spiritual, and metaphysical books in my early twenties, not long after I moved from New Jersey to California to try to find my way in the world. Before that move, I had no idea such books even existed.

And honestly, were it not for such books and my intense desire to learn, to grow, and to improve myself and my circumstances, I would have gone down a completely different road in life—a road I would rather not even think about or imagine.

Who could deny that books can and do change lives? It is my mission to write some of those books that do indeed change lives. I want people's lives to be better because I lived and because I wrote.

There are reasons I came into this life, and writing is one of them. I am living the life I was meant to live, and it is my sincere desire that you will live the life you were meant to live.

May I ask two favors of you? First, if you think this or any of my other books can help people in some of the ways they could use help, will you help spread the word about me and my writings? You could do that by loaning my books to others, giving my books as gifts, and by telling people about my books and about me. By doing these things, you will bless me beyond measure, and I truly believe you will bless others beyond measure as well.

Second, please consider writing an honest review for this book. Reviews are important to the success of any book and any author. And reviews really do help people decide whether or not a certain book is right for them. So, by writing a review, you will be helping me personally and other people as well.

And speaking of those other people: Say your review is the one that causes a person to actually buy this or any of my other books. And suppose that person then reads the book. And suppose that book helps that person to substantially improve their own life and the lives of others. Just imagine the possibilities—lives made better because you wrote a book review.

Currently, Amazon.com is the most important place you could post a review but feel free to post a review anywhere you choose. And keep in mind that even just a sentence or two could be sufficient. The number of words in a review you write is less important than what those words say.

And a new feature on Amazon is that you can now rate a book without even having to write a review.

You just click on the star rating of your choice and that's it. One click! So even if you are not going to write a review, you can easily have your voice heard.

And finally, always remember, you are capable of so much more than you have ever imagined. Learn, believe, act, and persist. If you will do those four things, nothing will stop you from continuing to build a better and better life for yourself and for those you care about.

Peace & Plenty . . .

ABOUT THE AUTHOR

James Goi Jr., aka The Attract Money Guru™, is the best-selling author of the internationally published *How to Attract Money Using Mind Power*, a book that set a new standard for concise, no-nonsense, straight-to-the-point self-help books. First published in 2007, that game-changing book continues to transform lives around the world. And though it would be years before James would write new books, and even more years before he would publish new books, that first book set the tone for his writing career. The tagline for James as an author and publisher is Books to Awaken, Uplift, and Empower™. And James takes those words seriously, as is evident in every book he writes. James: is a relative recluse and spends most of his time alone; is an advanced mind-power practitioner, a natural-born astral traveler, and an experienced lucid dreamer; has had life-changing encounters with both angels and demons, and even sees some dead people; has been the grateful recipient of an inordinate amount of life-saving divine intervention; is a poet and songwriter; is a genuinely nice guy who cares about people and all forms of life; fasts regularly; is a sincere seeker of higher human health; is an objective observer, a persistent ponderer, and a deliberate deducer; and has a supple sense of heady humor.

STAY IN TOUCH WITH JAMES

If you are a sincere seeker of spiritual truth and/or a determined pursuer of material wealth and success, James could be the lifeline and the go-to resource you have been hoping to find. Step One, subscribe to James's free monthly *Mind Power & Money Ezine* here: jamesgoijr.com/subscriber-page.html. Step Two, connect with James online anywhere and everywhere you can find him. You can start here:

Facebook.com/JamesGoiJr
Facebook.com/JamesGoiJrPublicPage
Facbook.com/HowToAttractMoneyUsingMindPower
Twitter.com/JamesGoiJr
Linkedin.com/in/JamesGoiJr
Pinterest.com/JamesGoiJr
Youtube.com/JamesGoiJr
Instagram.com/JamesGoiJr
Goodreads.com/JamesGoiJr
jamesgoijr.tumblr.com

James's Amazon Author Page

A great resource to help you keep abreast of James's ever-expanding list of books is his author page at Amazon.com. There you will find all of his published writings and have easy access to them in the various editions in which they will be published.

SPECIAL ACKNOWLEDGEMENT

To Kathy Darlene Hunt, who has been my rock, my Light, my safety net, and my buffer since I was in my twenties. She rightfully shares in the credit for every book I've written, for the books I'm working on now, and for every single book I will ever write.

Kathy Darlene Hunt
Author of *A Child of the Light*

A FREE GIFT FOR YOU!

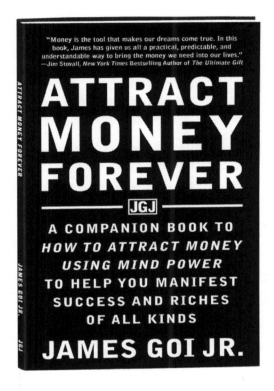

Attract Money Forever will deepen your understanding of metaphysics and mind-power principles as they relate to attracting money, manifesting abundance, and governing material reality. You'll learn how to use time-tested, time-honored, practical, and spiritual techniques to be more prosperous and improve your life in astounding and meaningful ways. Visit jamesgoijr.com/subscriber-page.html for your free download copy of this amazing book and to receive James's free monthly *Mind Power & Money Ezine*.

FURTHER READING

The 80/20 Principle by Richard Koch
The ABCs of Success by Bob Proctor
Abundance Now by Lisa Nichols and Janet Switzer
Act Like a Success, Think Like a Success by Steve Harvey
The Amazing Power of Deliberate Intent by Esther Hicks and Jerry Hicks
As a Man Thinketh by James Allen
The Awakened Millionaire by Joe Vitale
Awaken the Giant Within by Tony Robbins
Being and Vibration by Joseph Rael with Mary Elizabeth Marlow
The Biology of Belief by Bruce H. Lipton, Ph.D.
Breaking the Habit of Being Yourself by Dr. Joe Dispenza
The Charge by Brendon Burchard
Choice Point by Harry Massey and David R. Hamilton, Ph.D.
Clarity by Jamie Smart
The Compound Effect by Darren Hardy
The Cosmic Code by Heinz R. Pagels
The Cosmic Ordering Service by Barbel Mohr
The Council of Light by Danielle Rama Hoffman
Create Your Own Future by Brian Tracy
Creating on Purpose by Anodea Judith and Lion Goodman
Creative Visualization by Shakti Gawain
The Dancing Wu Li Masters by Gary Zukav
The Diamond in Your Pocket by Gangaji
The Dice Game of Shiva by Richard Smoley
Divine Audacity by Linda Martella-Whitsett
The Divine Matrix by Gregg Braden
Dreamed Up Reality by Dr. Bernardo Kastrup

The Dynamic Laws of Prosperity by Catherine Ponder

Emergence by Derek Rydall

Feeling Is the Secret by Neville Goddard

The Field by Lynne McTaggart

Follow Your Passion, Find Your Power by Bob Doyle

The Four Desires by Rod Stryker

Frequency by Penney Peirce

The Game of Life and How to Play It by Florence Scovel Shinn

Having It All by John Assaraf

The Hidden Power by Thomas Troward

How Consciousness Commands Matter by Dr. Larry Farwell

How Successful People Think by John C. Maxwell

I AM by Vivian E. Amis

I Wish I Knew This 20 Years Ago by Justin Perry

Infinite Potential by Lothar Schafer

Instant Motivation by Chantal Burns

It Works by RHJ

Jack Canfield's Key to Living the Law of Attraction by Jack Canfield and D.D. Watkins

Just Ask the Universe by Michael Samuels

Key to Yourself by Venice J. Bloodworth

The Law of Agreement by Tony Burroghs

Lessons in Truth by H. Emilie Cady

Life Power and How to Use It by Elizabeth Towne

Life Visioning by Michael Bernard Beckwith

Live Your Dreams by Les Brown

The Lost Writings of Wu Hsin by Wu Hsin and Roy Melvyn (Translator)

The Magical Approach by Seth, Jane Roberts, and Robert F. Butts

The Magic Lamp by Keith Ellis

The Magic of Believing by Claude M. Bristol

The Magic of Thinking Big by David J. Schwartz

Make Magic of Your Life by T. Thorne Coyle
Manifesting Change by Mike Dooley
The Map by Boni LonnsBurry
The Master Key System by Charles F. Haanel
The Millionaire Mind by Thomas J. Stanley
Mind and Success by W. Ellis Williams
Mind into Matter by Fred Alan Wolf, Ph.D.
Mind Power into the 21st Century by John Kehoe
Miracles by Stuart Wilde
The Miracles in You by Mark Victor Hansen and Ben Carson (Foreword)
Mysticism and the New Physics by Michael Talbot
New Physics and the Mind by Robert Paster
The One Command by Asara Lovejoy
One Mind by Larry Dossey, M.D.
The One Thing by Garry Keller with Jay Papasan
One Simple Idea by Mitch Horowitz
Our Invisible Supply by Frances Larimer Warner
Our Wishes Fulfilled by Dr. Wayne W. Dyer
Physics on the Fringe by Margaret Wertheim
Playing the Quantum Field by Brenda Anderson
The Power of Now by Eckhart Tolle
The Power of Positive Thinking by Dr. Norman Vincent Peale
The Power of Your Subconscious Mind by Joseph Murphy
Power through Constructive Thinking by Emmet Fox
The Power to Get Things Done by Steve Levinson Ph.D. and Chris Cooper
Programming the Universe by Seth Lloyd
Prosperity by Charles Fillmore
Psycho-Cybernetics by Maxwell Maltz
Quantum Creativity by Pamela Meyer
Quantum Reality by Nick Herbert
The Quantum Self by Danah Zohar

Reality Creation 101 by Christopher A. Pinckley
Reality Unveiled by Ziad Masri
The Sacred Six by JB Glossinger
The School of Greatness by Lewis Howes
The Science of Getting Rich by Wallace D. Wattles
The Science of Mind by Ernest Holmes
The Secret by Rhonda Byrne
The Secret of the Ages by Robert Collier
The Self-Aware Universe by Amit Goswami
Shadows of the Mind by Roger Penrose
Shift Your Mind by Steve Chandler
The Slight Edge by Jeff Olson
Soul Purpose by Mark Thurstan, Ph.D.
Spiritual Economics by Eric Butterworth
Supreme Influence by Niurka
There Are No Accidents by Robert E. Hopcke
Think and Grow Rich by Napoleon Hill
Thought Power by Annie Besant
Thoughts Are Things by Prentice Mulford
True Purpose by Tim Kelley
The Universe Is a Dream by Alexander Marchand
Unleash Your Full Potential by James Rick
The Way of Liberation by Adyashanti
What Is Self? by Bernadette Roberts
The Wisdom Within by Dr. Irving Oyle and Susan Jean
Within the Power of Universal Mind by Rochelle Sparrow and Courtney Kane
Working with The Law by Raymond Holliwell
You Are the Universe by Deepak Chopra and Menas C. Kafatos
You Are the World by Jiddu Krishnamurti
Your Invisible Power by Genevieve Behrend
You Unlimited by Norman S. Lunde
The Zigzag Principle by Rich Christiansen

Made in the USA
Coppell, TX
27 November 2022

87195558R00037